Sprinkles
the Fire Dog

Sprinkles
the Fire Dog

Written by

**Frank
Viscuso**

Illustrated by

**Paul
Combs**

Copyright © 2021 by
Fire Engineering Books & Videos
110 S. Hartford Ave., Suite 200
Tulsa, Oklahoma 74120 USA

800.752.9764
+1.918.831.9421
info@fireengineeringbooks.com
www.FireEngineeringBooks.com

Library of Congress Cataloging-in-Publication Data

Names: Viscuso, Frank, author. | Combs, Paul (Paul David), 1966-
    illustrator.
Title: Sprinkles the fire dog / written by Frank Viscuso ; illustrated by
    Paul Combs.
Description: Tulsa, Oklahoma : Fire Engineering Books & Videos, [2021] |
    Audience: Grades K-1. | Summary: Teased by the other dogs for being
    small and wanting to work at the fire station, a young pup learns that
    if he trains hard and believes in himself, he can do anything, even
    become a fire dog.
Identifiers: LCCN 2021021488 | ISBN 9781593705077 (hardcover)
Subjects: CYAC: Dogs--Fiction. | Self-confidence--Fiction. | Fire
    fighters--Fiction.
Classification: LCC PZ7.1.V58 Sp 2021 | DDC [E]--dc23
LC record available at https://lccn.loc.gov/2021021488

Printed in the United States of America

1 2 3 4 5   25 24 23 22 21

I could not express enough gratitude to my friend Paul Combs for bringing life to this story through his incredible artwork. Paul's talent is a gift from above, and what he chooses to do with it is a gift to all of us. Thank you, Brother, for everything you have done, and will continue to do, to help make this world a better place.

Paul and I are both thankful to partner with our team at Clarion Events and Fire Engineering Books & Videos on our first children's book, specifically, Diane Rothschild, Joshua Neal, Mark Haugh, Holly Fournier, Christopher Barton, Tony Quinn, and Bobby Halton. You have always supported our vision and ideas, and we are honored to be part of this team.

To my wife Laura, who pulled this story out of a time capsule and asked me to read it to my boys Frankie and Nicholas, thank you for encouraging me to finally introduce Sprinkles the Fire Dog to the world.

I would like to dedicate this book to Frankie, Nicholas, and all the other young boys and girls out there who are setting goals, putting in the work, and turning their dreams into reality.

—Frank

It is not often that you get a chance in life to create something extraordinary, to pull inspiration from the imagination storm that rages inside each artist and channel the creative lighting bolts through your hand and onto paper. It is these moments that artists live for!

It is this very opportunity that my friend Frank presented to me in the way of a small dog who has very, very big dreams—a story that we all can relate to. It is with great pride and humility that I thank him for trusting me to bring his vision to life, and for his patience while I overcame my own mutts on the corner (it's in the book). Frank, from the bottom of my creative soul I say thank you, and I can't wait to see how Sprinkles inspires so many others as he has inspired me.

To my wife, Sheryl, and daughter, Brittney—as always, you are my rock and my reason for striving for whatever greatness I can achieve. You are my inspiration—forever and always!

—Paul

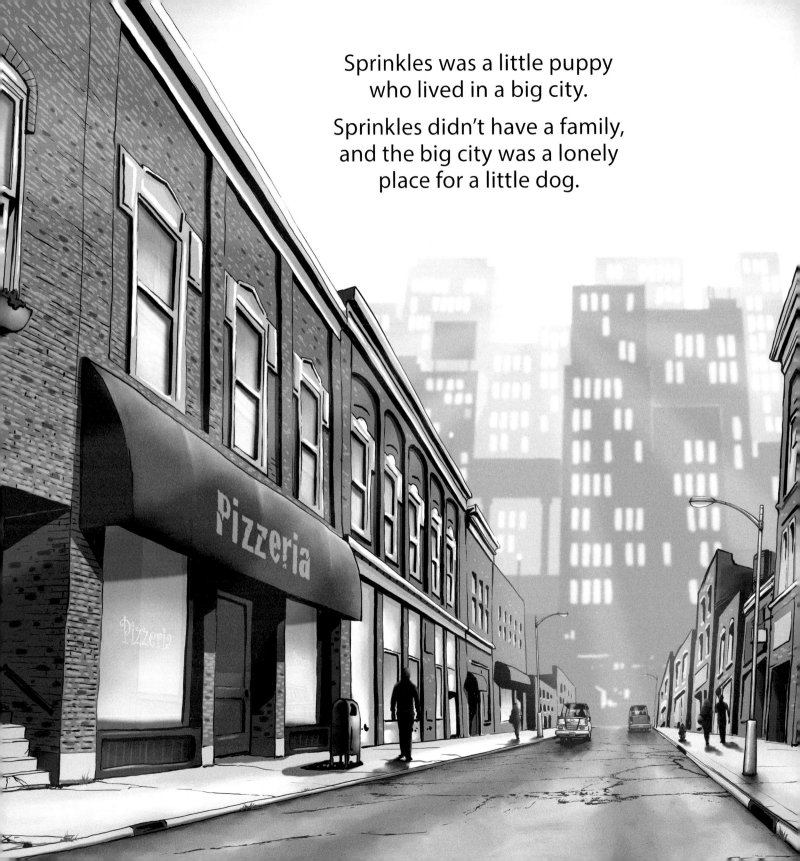

Sprinkles was a little puppy
who lived in a big city.

Sprinkles didn't have a family,
and the big city was a lonely
place for a little dog.

He tried to make friends with other city dogs, but they laughed at him because he was so small, and they were all so big.

Plug the pug and the other mutts never left their street corner. They just sat around all day, every day, and made fun of all the other dogs and cats.

They teased Sprinkles the most because he was a dalmatian with only a few small spots. "You're not even a dalmatian!" Plug laughed. "Dalmatians are big and strong and have many spots. You're tiny and you only have a few."

"And you can't even bark," one of the mutts added, as Sprinkles walked away with his head hanging low.

One day, Sprinkles was walking down the street when he heard the sound of loud sirens coming closer and closer.

The sirens scared Sprinkles, so he jumped behind a trash can and peeked his head out to see what was making the noise.

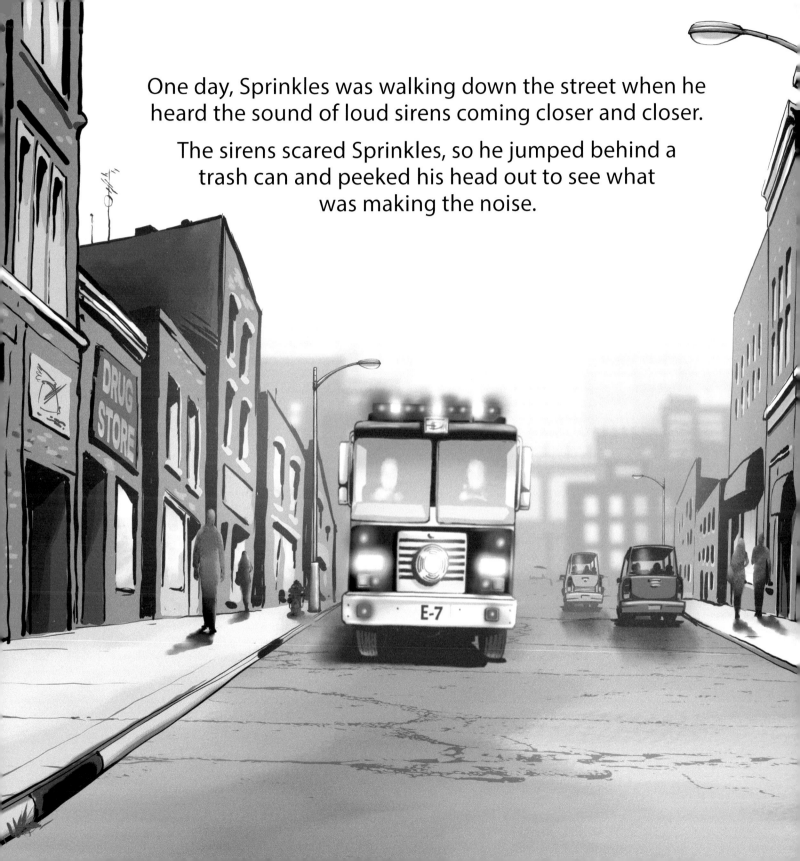

That's when he saw the grandest sight his eyes had ever seen—a huge red fire engine. Riding on the fire engine were four brave firefighters and a big, strong dalmatian.

At that moment, Sprinkles knew what he wanted to be—a fire dog.

So, Sprinkles went down
to the firehouse.

He was very nervous because he
wanted the firefighters to like him.

He looked around but there
was nobody in sight—
no firefighters,
no dalmatian,
only the big red fire engine.

Sprinkles looked at the fire engine with wide eyes. He wondered what it would be like to ride on it. He imagined sitting in the front seat, with his head out of the window, racing to a fire with the firefighters to go help people.

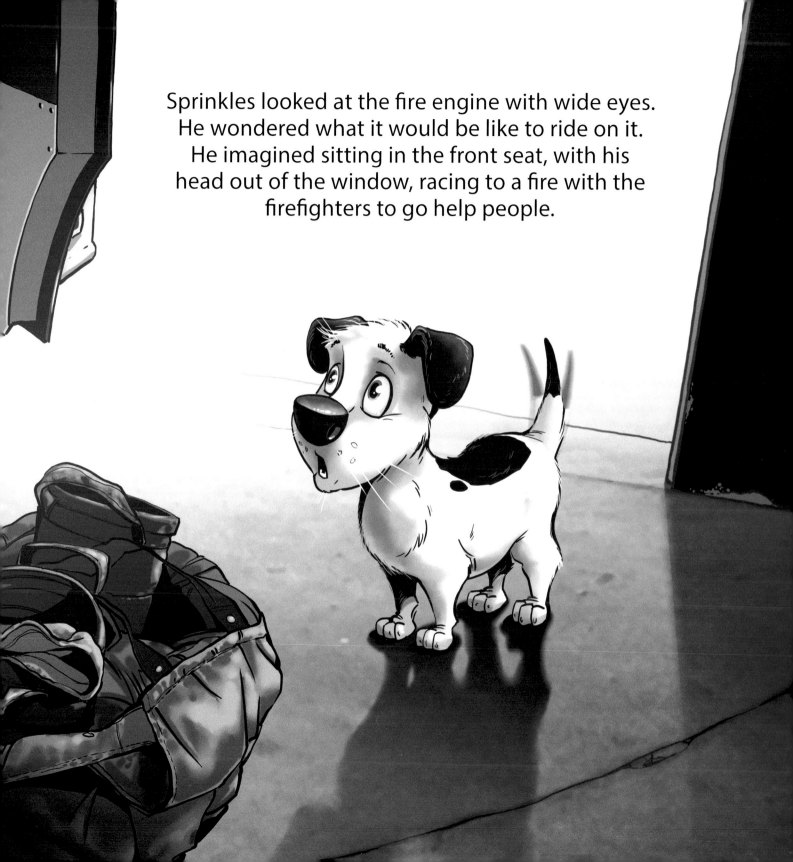

Suddenly, loud bells started to ring. Sprinkles tried to hide, but there was no place to go. Out of nowhere, four firefighters came sliding down a pole and ran to the fire engine.

Sprinkles tried to move out of the way, but they were coming from every direction. One firefighter almost stepped on him.

Another firefighter named Frankie looked at Sprinkles and said, "Be careful, little pup, you might get hurt."

Just then, the big, strong dalmatian named Ember ran
by Sprinkles and jumped onto the fire engine, and off they
went with their flashing lights and loud siren leading the way.

Sprinkles just stood there and
watched until the fire engine
was out of sight.

Across the street, Plug and the mutts were laughing, "See, we told you that you couldn't be a fire dog."

All three of them continued to tease Sprinkles as he lowered his head and walked away. He felt sad and lonely.

Later that day, Sprinkles was lying down
on his own corner when he spotted the
fire engine driving down the street
in his direction.

He quickly sat up to watch them pass, but
instead, they stopped right in front of him.

Sprinkles was super excited. He thought they were
stopping to see him, but he soon found out they just
needed to refill their fire engine tank with water from
the fire hydrant.

Sprinkles became sad again. He lowered his shoulders
and began to walk away.

"Hey, isn't that the little pup who was at our firehouse?"
one firefighter asked.

"Yes, that's him," Frankie answered.
"I almost stepped on him."

"Maybe we should take him back to the firehouse with us.
We could use another fire dog."

"No way," said the Captain. "He's too small.
He'll only get in the way."

Sprinkles knew he had to prove himself, so he took
a deep breath and tried to bark as loud and strong
as he could, but when he tried, nothing came out.

The Captain turned to Frankie and said,
"You see, he wouldn't make a good fire dog.
He can't even bark."

The firefighters and their big, strong dalmatian jumped back on the fire engine. Frankie knelt down to pet Sprinkles and said, "Don't worry, little fella, one day you'll be a little older and a lot stronger like Ember, and when that day comes, you will be a great fire dog."

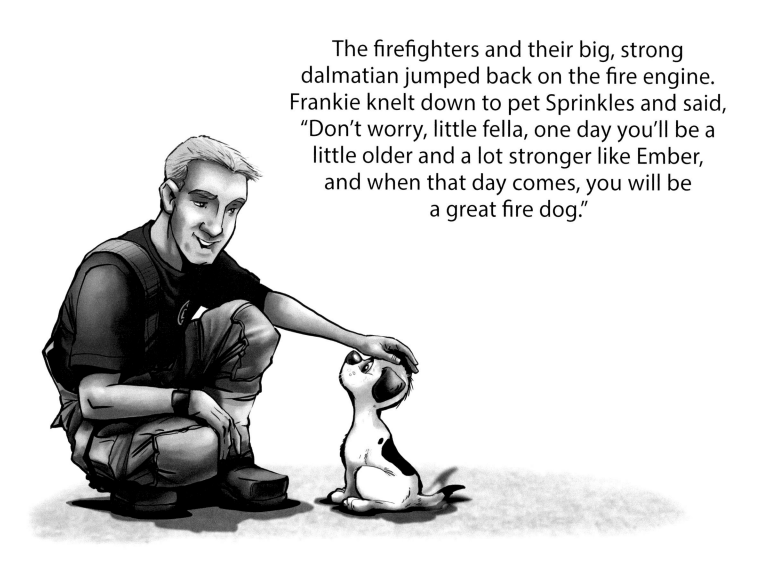

Sprinkles was so sad. He really wanted to be a fire dog, but he was starting to think it was all just a silly dream that was not going to come true. After all, the corner dogs always laughed at him, and the firefighters only saw a little puppy when they looked at him.

The next day, Sprinkles walked around the big city feeling sorry for himself. He knew he had a stronger bark inside of him, but he missed his chance to prove it. He started to believe what the corner dogs were saying about him.

Maybe it was just a silly little dream.

But then it came to him: The firefighters did not say "no" forever. They only said "no" for now.

Sprinkles knew what he had to do. He decided that he was going to work harder than ever before.

Every day, he ran as fast as he could and practiced his bark so he would be ready for the next time he had a chance to prove himself.

A few months later, he was running so fast that he could no longer hear the corner dogs laughing when he passed them.

One night, Sprinkles was walking down the street when he smelled smoke. He looked up to see fire coming from the window of a nearby house.

He looked for someone to help, but there was no one around.

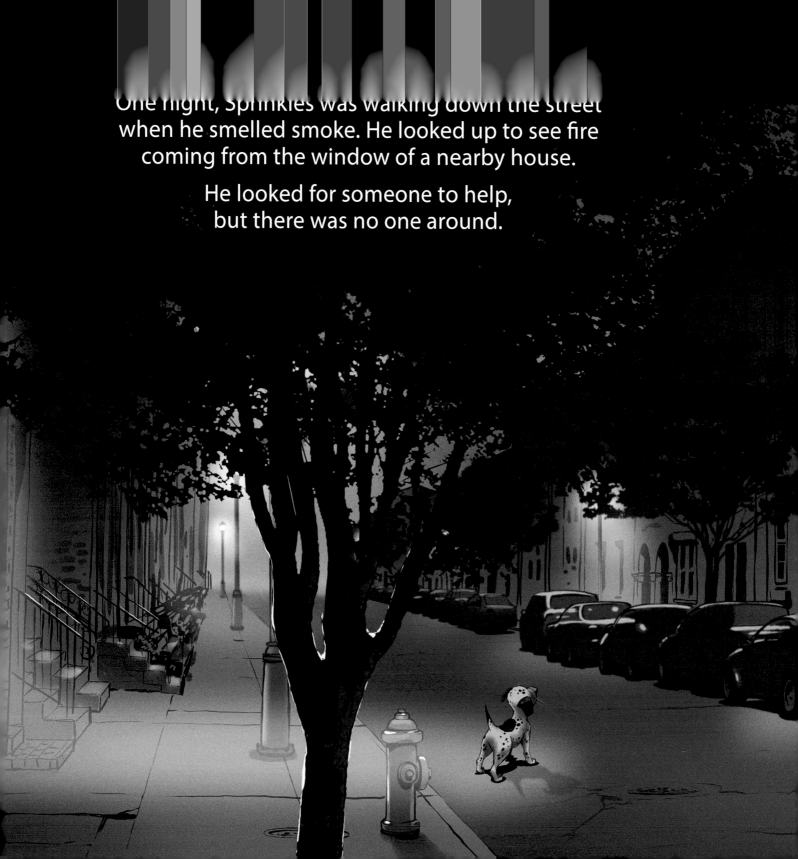

He knew he had to do something, so he took a deep breath.
Then he barked as loud as he could.

His bark was so loud that even he couldn't believe it. He kept
barking until everyone knew there was danger in the area.

Suddenly, a man looked out of his window to see what was happening and realized his house was on fire.

The man ran outside, rushed up to Sprinkles, and said, "Thank you, little puppy. You saved my life!"

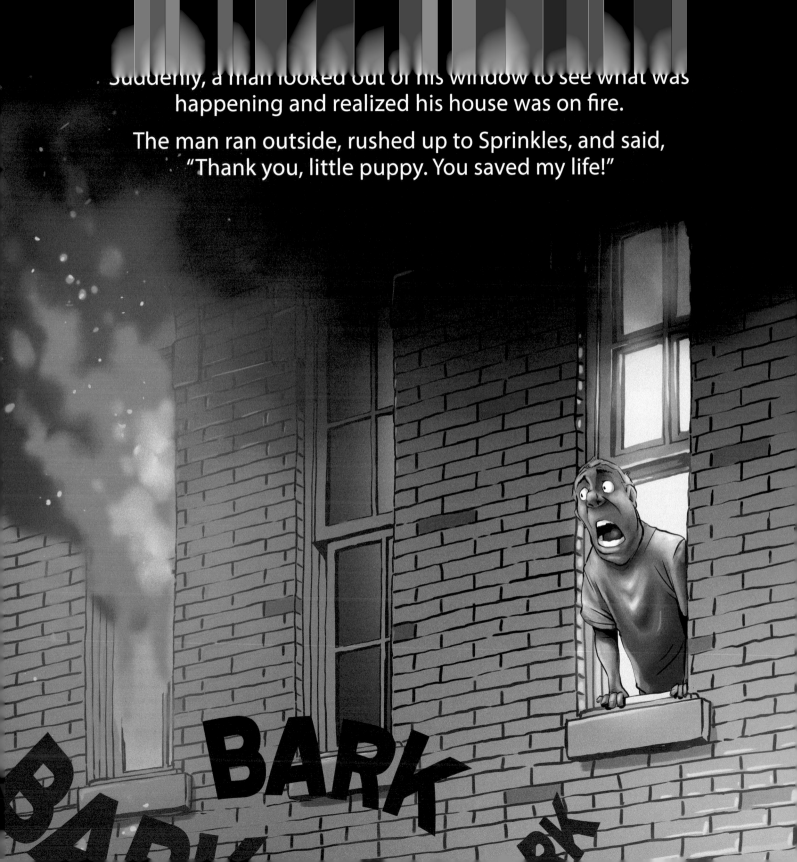

But Sprinkles was not finished yet.

He ran through the city streets as fast as he could,
faster than he had ever run before.

He ran past Plug and the mutts all the way down to the
firehouse and started barking again—this time even louder!

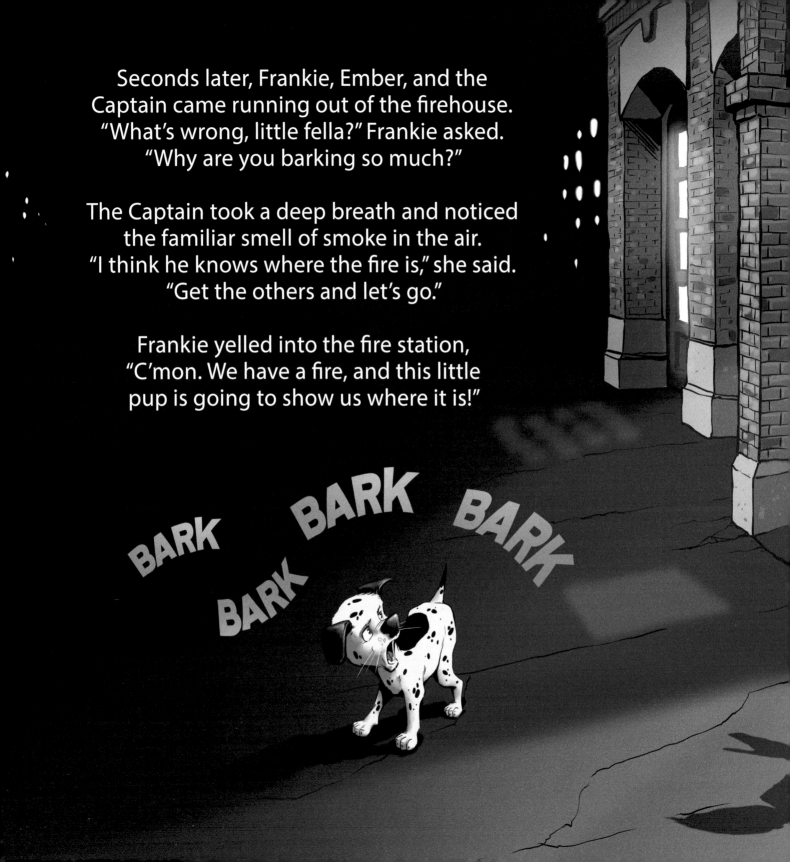

Seconds later, Frankie, Ember, and the
Captain came running out of the firehouse.
"What's wrong, little fella?" Frankie asked.
"Why are you barking so much?"

The Captain took a deep breath and noticed
the familiar smell of smoke in the air.
"I think he knows where the fire is," she said.
"Get the others and let's go."

Frankie yelled into the fire station,
"C'mon. We have a fire, and this little
pup is going to show us where it is!"

BARK BARK BARK
BARK BARK

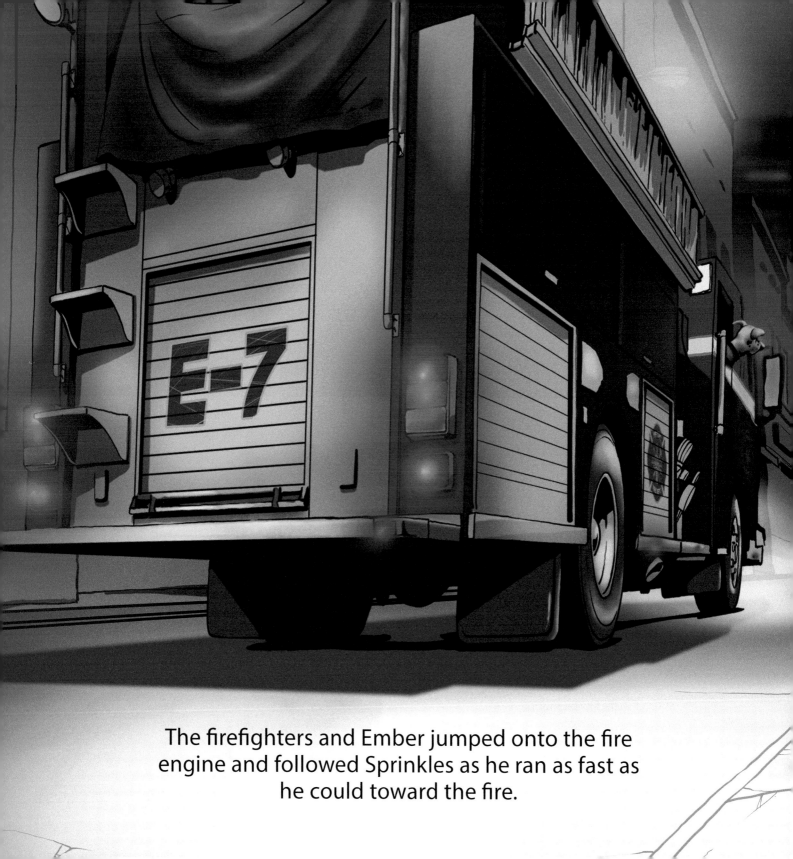

The firefighters and Ember jumped onto the fire engine and followed Sprinkles as he ran as fast as he could toward the fire.

Plug and the mutts watched as Sprinkles ran by.
They were amazed to see the fire engine behind him.
So, they left their comfortable street corner for
the first time in years to see what was going on.

Sprinkles led the fire engine straight to the fire. Once the firefighters saw the smoke, they quickly went to work and put the fire out.

After the fire, the man who lived in the house told the firefighters how Sprinkles saved his life.

The firefighters gathered around Sprinkles.
"Little buddy, I was wrong about you,"
the Captain said. "If you still want to be a fire dog,
we would love to have you on our team."

Sprinkles barked proudly as the firefighters praised him for the job he had done.

Frankie leaned over and said, "I knew you could do it, buddy. I'm proud of you."

BARK BARK

The firefighters, Ember, and Sprinkles drove away
on the big red fire engine.

Sprinkles never gave up and he never let the other dogs take
his dream away from him. He worked hard and believed in
himself, and his dream came true...

and now he is a fire dog!

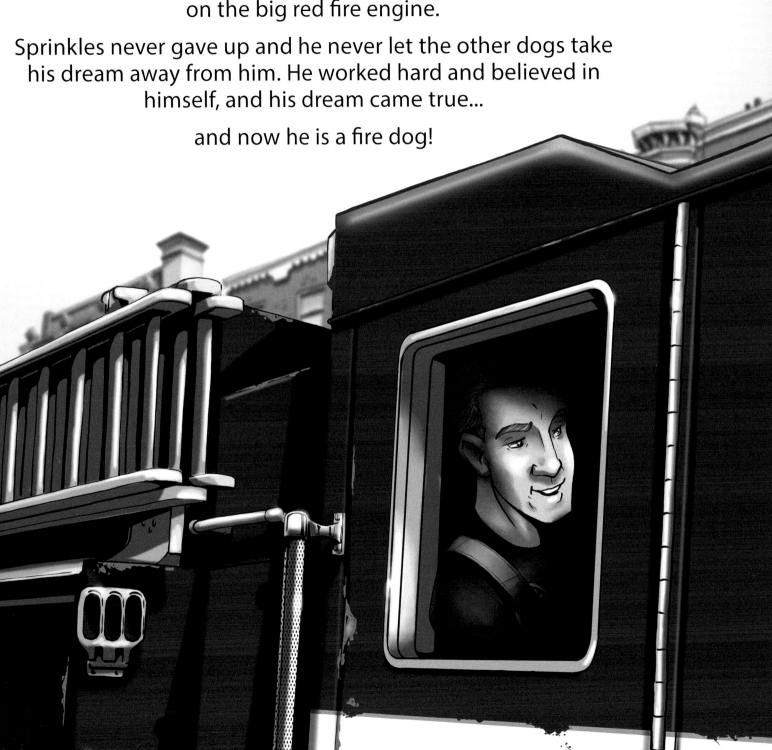

Questions for the Young Reader

Why was Sprinkles so sad at the beginning of the story?

What would happen if Sprinkles believed what the corner mutts were saying about him?

If Sprinkles did not work hard and believe in himself, do you think he would have become a fire dog?

What are some of your dreams?

What can you start doing to help make your dreams come true?

**Be kind and be brave.
You can do anything if you work hard
and believe in yourself.**